S.O.S

Rewiring Your Sense-Of-Self with

CREATIVE INTELLIGENCE TRAINING

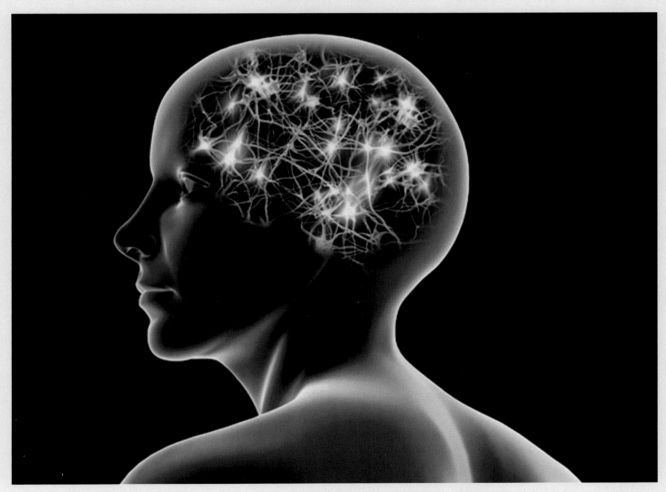

Neural Networks, Pasieka, Getty Images

Logosoma Living Wellness Protocol
And Psychedelic Medicine

Phillip Romero, MD

To order additional copies of this book, contact:
Xlibris
844-714-8691
www.Xlibris.com
Orders@Xlibris.com

Scripture quotations marked KJV are from the Holy Bible, King James Version (Authorized Version). First published in 1611. Quoted from the KJV Classic Reference Bible, Copyright © 1983 by The Zondervan Corporation.

Library of Congress Control Number:		2022904782
ISBN:	Softcover	978-1-6698-1518-1
	Hardcover	978-1-6698-1519-8
	EBook	978-1-6698-1517-4

Print information available on the last page

Rev. date: 10/03/2022

DEDICATED TO MY BELOVED DAUGHTER SAYUME ANN

It is not the strongest or the most intelligent who will survive but those who can best manage change.

Charles Darwin

Intelligence is the ability to adapt to change.

Stephen Hawking

The view of the brain as a creativity machine that constantly uses inferences and guesses to reconstruct the external world... was a dramatic shift...

Eric R. Kandel in

The Age of Insight: The Quest to Understand the Unconscious in Art, Mind, and Brain

S.O.S.

- *Self-organizing systems*
- *Sense-of-Self*
- *Save Our Ship*

SOS marks the meaning of many things.

Rendering connectivity,

Illuminating conscious awareness, and

Signaling distress,

SOS maps the vitality of life,

Revealing a perpetual dynamic of

Creative and Destructive forces.

We live in a web of complex self-organizing systems

From the cosmic dance between the classic Newtonian universe and

The quantum substrate of all existence

To the biosphere of our planet.

Our sense-of-self is connected to people, places, and things

Through infinite networks, bridges, and bonds.

The links are physical, emotional, conscious, and non-conscious,

Charged with never-ending change,

And triggering the Anxiety of Impermanence.

The sustainable wellness of any system requires balancing the

Forces that drive life.

CONTENTS

INTRODUCTION

I

THE LOGOSOMA LIVING WELLNESS PROTOCOL

II

LOGOSOMA BRAIN TRAINING

III

CREATIVE INTELLIGENCE TRAINNG

IV
PSYCHOPOIESIS TRAINING

V
CHANGE THE SENSE-OF-SELF;
REMAKE THE WORLD

The Universe is a Creativity Machine.

The universe is a creativity machine.

Autopoiesis, self-making, charges each moment in the

Billions of years of cosmic development.

From the Nothing before the Big Bang,

The self-creating dynamics of Nature

Drive simplicity toward complexity,

From darkness into light,

From stillness to perpetual motion.

From emptiness into form,

And from form into emptiness.

I emerge as a living system,

Distinguishing myself from others,

Through the bodies of my parents,

Into a body of my own,

Empowered with psychopoiesis,

The drive to manifest a sense-of-self,

Creating a perpetual narrative of meaningful connections
With others.

With gratitude and respect for the nature of Nature's creativity,

I devote myself to the realization of the full potential of my being in
action.

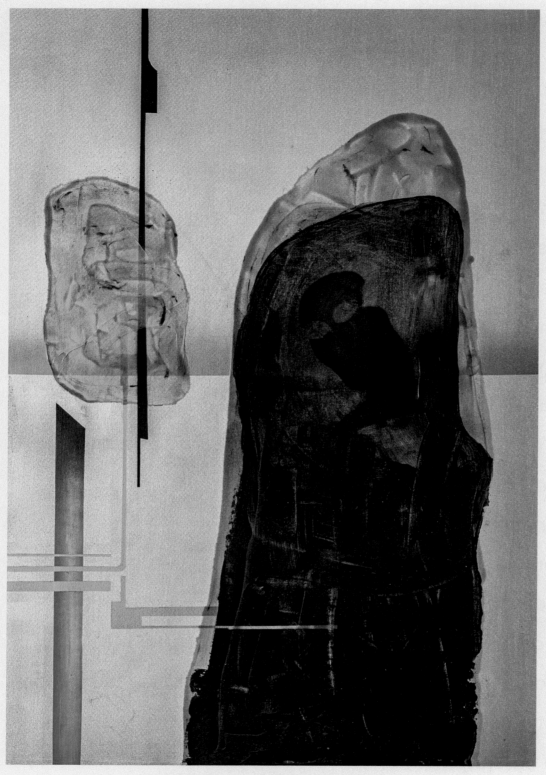

ARTIFICIAL ANNUNCIATION

(After Leonardo). 1976, 48" x 60" Oil/Canvas
Phillip Romero (Photographed by Jennifer Douglas)

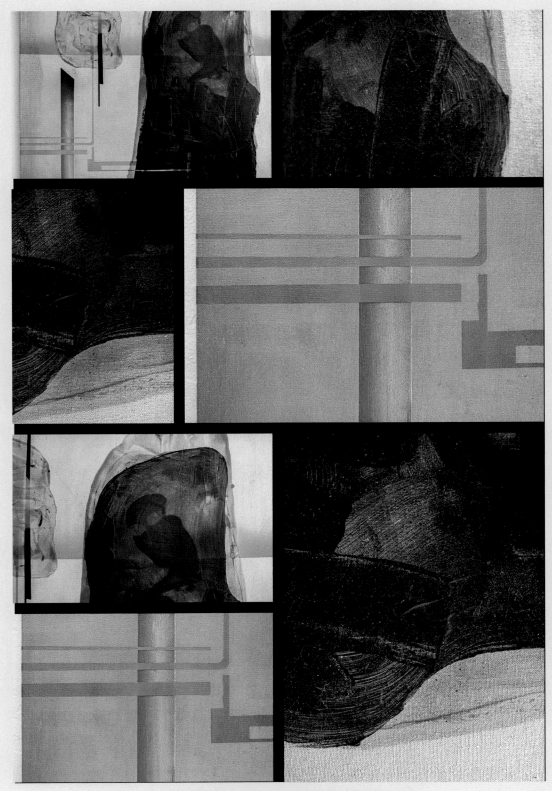

ARTIFICIAL ANNUNCIATION

(After Leonardo). 1976, 48" x 60" Oil/Canvas
Phillip Romero (Photographed by Jennifer Douglas)

Introduction

Reinventing Ourselves

YUME: DREAM, Phillip Romero, 2022

Creative Intelligence: *The Human Super-Power*

Human beings are endowed with a potential superpower—*Creative Intelligence*. Creative Intelligence (CI) is critical to human survival, and it drives human cultural evolution. CI is defined here as the *mindful coordination* of the genetically designed traits of *resilience* to adversity and the *creative* problem-solving skills encoded in DNA and hardwired in the neural networks of the brain. The brain is a creativity machine. Mindfulness training exponentially amplifies human Creative Intelligence. *Resilience + Creativity x Mindful Coordination = Creative Intelligence.* $R + C \ x \ MC=CI$.

Like emotional intelligence, Creative Intelligence can be cultivated through learning. Creative Intelligence Training (CIT) empowers the brain-mind system to continually reinvent the *Sense-Of-Self* (SOS) to adapt to the stresses of life. The core of all human adversity, the 'anxiety of impermanence,' can mastered with Creative Intelligence.

Logosoma Living Wellness Protocol (LLWP) empowers Creative Intelligence to mitigate relationship stress, to rewire the neural substrates of the sense-of-self that trigger self-sabotaging attitudes and behaviors, and to integrate psychedelic medicine to promote the process.

Anxiety of Impermanence: *Attachment Theory, Ancient and Modern*

Buddhism identified the fear of the loss of attachment as the core of human suffering. Developmental neuroscience reveals that insecure attachment plays a critical role in existential anxiety that undermines an authentic sense-of-self. The fact of impermanence is a significant cause of stress related physical, mental, emotional, and relationship dysfunction. Vincent Felitti MD (1998) demonstrated that Adverse Childhood Experience (ACE) can have life-long consequences on health and longevity.

Mastering the existential fear of impermanence is the primary focus of the Logosoma Living Wellness Protocol. Ancient practices have been shown to reduce the egoistic clinging to the self and liberate the mind-body into a deeper sense of connectivity. Neuroscience is beginning to reveal the brain mechanisms for this experience and the wellness benefits that come with regular practice.

Reinventing Ourselves: *The Logosoma Living Wellness Protocol*

SOS: Rewiring the Sense-of-Self with Creative Intelligence Training integrates clinical neuroscience, mindfulness philosophy and practice, and psychedelic medicine. The *Logosoma Living Wellness Protocol* (LLWP) recruits the brain-mind network that endows human beings with the power to continuously recreate themselves. Cultural evolution is a story of perpetual reshaping the patterns of how we live with each other. The challenge for each generation is to creatively transform the negative patterns of self-sabotage that arise in social systems from families to society.

LLWP presents a three-module curriculum. The first module, Logosoma Brain Training (LBT), is a clinical methodology I developed in the 1980s for mitigating relationship stress and recruiting innate individual creativity. The themes of Self-Creativity (autopoiesis) and Resilience to adversity were presented in *Phantom Stress:*

Brain Training to Master Relationship Stress (2010) and *The Art Imperative: The Secret Power of Art* (2010).

The second module, Creative Intelligence Training (CIT), extends LBT to focus on recreating the sense-of-self. I presented this methodology in *Global Village Burnout: Creative Intelligence = Survival* (2021). The neuroscience of creativity reveals the creative potential for any person can be improved with mindfulness training.

The third module, Psychopoiesis Training (PT), uses psychedelic medication to activate neuroplasticity. New science of psychedelics offers the potential to rapidly kindle neuroplasticity with mindfulness practices for long term changes. Psychedelic medicine demonstrates marked efficacy with treatment resistant depression (TRD), posttraumatic stress (PTSD), addiction, and anxiety spectrum disorders.

As an artist, a psychiatrist, a mindfulness practitioner, and author, I comingled many domains of knowledge into a consilient wellness protocol (consilience = unity of knowledge).

Mastering Stress with Consilience: *The Unity of Knowledge*

SOS is inspired by E.O. Wilson, evolutionary biologist, who wrote *Consilience: The Unity of Knowledge (1998)* and neuroscientist Bruce McEwen, who wrote *The Hostage Brain,* (1994), and *The End of Stress as We Know It,* (2002). Ed Wilson and Bruce McEwen organized a three-day Consilience conference with the New York Academy of Science and Rockefeller University 1999. Bruce invited me to many conferences on the neuroscience of stress for over twenty years. With his supervision, I was able to "put the science" in

Logosoma Brain Training to make it a fact-based methodology, and he wrote the cover comment for *Phantom Stress.* Ed encouraged my book, *The Art Imperative*, and wrote a cover comment. My career path has been devoted to integrating three domains of knowledge, creativity, consciousness, and medicine, into Logosoma Living Wellness Protocol.

As Assistant Professor of Child Psychiatry at New York-Presbyterian Hospital-Weill Medical College, Cornell University, the Logosoma paradigm guided supervision for Child Psychiatry Fellows in a course on family systems for 25 years. In my private practice of family psychiatry, LBT and CIT have been clinically applied for over 40,000 hours in direct patient care and 5,000 hours of physician and support staff training.

Consciousness: *Cultivating Glimpses of Selfless Mind*

Ancient theories of consciousness identify the idea of awareness without a sense-of-self—*selfless mind.* This concept of cosmic consciousness has some biological grounding in that all living self-organizing-systems have some sensory ability in the cytoskeletal structures to distinguish the intracellular realm from the extra-cellular world with the cell wall—a rudimentary distinction between a sense-of-self and not-self. Buddhist philosophy identifies the self-centered clinging ego as the root of human suffering. The Eightfold Path to enlightenment and liberation requires embracing the fact that everything is impermanent. Cultivating mindfulness is critical to understand one's perception of reality is relative.

With years of rigorous mindfulness training, experiencing consciousness beyond the egoistic sense-of-self, *selfless mind,* is possible. Opening one's perceptions to the vital forces of nature's creativity mitigates the fear of loss and dying—*the anxiety of impermanence*. Many cultures used medicinal psychedelic plants to 'glimpse' the selfless mind, transcending the sense-of-self to explore the complex interconnectivity and attain peace with living.

Brain Scan #1: *The Neuroscience of Mindfulness Meditation,* Yi-Yuan Tang, Britta K. Hölzel, and Michael I. Posner review the state of the art of research. (NATURE REVIEWS | NEUROSCIENCE VOLUME 16 | APRIL 2015 | 225 © 2015 Macmillan Publishers Limited)

Mindfulness meditation

Different styles and forms of meditation are found in almost all cultures and religions. Mindfulness meditation originally stems from Buddhist meditation traditions. Since the 1990s, mindfulness meditation has been applied to multiple mental and physical health conditions and has received much attention in psychological research. In current clinical and research contexts, mindfulness meditation is typically described as non-judgmental attention to experiences in the present moment. This definition encompasses the Buddhist concepts of mindfulness and equanimity and describes practices that require both the regulation of attention (to maintain the focus on immediate experiences, such as thoughts, emotions, body posture and sensations) and the ability to approach one's experiences with openness and acceptance...

The mindfulness practices that have been the subject of neuroscientific research comprise a broad range of methods and techniques, including Buddhist meditation traditions, such as Vipassana meditation, Dzogchen, and Zen, as well as mindfulness-based approaches such as integrative body–mind training (IBMT), mindfulness-based stress reduction (MBSR) and clinical interventions based on MBSR…

…mindfulness meditation includes at least three components that interact closely to constitute a process of enhanced self-regulation: enhanced attention control, improved emotion regulation and altered self-awareness (diminished self-referential processing and enhanced body awareness).

Mindfulness and self-awareness

According to Buddhist philosophy, the identification with a static concept of 'self' causes psychological distress. Dis-identification from such a static self-concept result in the freedom to experience a more genuine way of being. Through enhanced meta-awareness (making awareness itself an object of attention), mindfulness meditation is thought to facilitate a detachment from identification with the self as a static entity and a tendency to identify with the phenomenon of 'experiencing' itself is said to emerge…

…there is emerging evidence that mindfulness meditation might cause neuroplastic changes in the structure and function of brain regions involved in regulation of attention, emotion, and self-awareness.

The Default Mode Network and the Sense-of-Self

…the default mode network (DMN) is involved in self-referential processing… These structures show high activity during rest, mind wandering, and conditions of stimulus-independent though, and have been suggested to support diverse mechanisms by which an individual can 'project' themselves into another perspective. fMRI studies have investigated activity in the DMN in association with mindfulness practice. Regions of the DMN (the medial PFC and PCC) showed relatively little activity in meditators compared to controls across different types of meditation, which has been interpreted as indicating diminished self-referential processing.

Psychopoiesis: *Recreating the Sense-of-Self with Psychedelics*

Psychopoiesis, the ability to activate the brain's neuroplasticity to renew the sense-of-self has been demonstrated with many types of life experiences, ancient mindfulness practices, philosophical contemplation, spiritual traditions, and psychotherapy.

In 2019 the FDA approved esketamine. Psychedelic medication is now integrated with *Logosoma Living Wellness Protocol*. With over thirty-five years of clinical treatment of depression, posttraumatic stress disorder, addiction, and obsessive-compulsive spectrum disorders, the long-awaited use of psychedelic medicine for promoting the brain's neuroplasticity to empower the LBT module for Selfless Mind.

Selfless mind refers to the mindful attainment of consciousness without egoism. Experiencing consciousness beyond the sense-of-self requires years of rigorous meditation, yoga, and creative practices.

First Contact with Psychedelics

I came of age in the psychedelic 1960s. My first job out of high school was as a firefighter while starting pre-med studies. A war veteran firefighter had a karate school where I earned a green belt. That year I started Transcendental meditation and began making paintings inspired by Picasso and Matisse. The following summer, I worked as a stage guard at the Woodstock Pop Festival and immersed myself in the massive wave of Pop culture. I discovered Andy Warhol and Robert Rauschenberg, a fellow Port Arthurian, as my new muses for art. My first exposure to LSD was watching many people have 'bad-trips' at the music concert. It inspired me to write a research paper on serotonin and LSD for my pre-med biochemistry course. As an intrepid college student, I took a high potency 'windowpane' gel-tab of Owsley LSD. These experiences led to a lifelong study of consciousness and creativity.

While in medical school, my studies took me to a summer studying Tibetan Medicine and Buddhism in Dharamsala, India, where I received private initiations with HH the Dalai Lama. I kept a watchful eye on psychedelic research throughout my medical career. I hoped that neuroscience would reveal the basis of the enlightenment experience. That time has come.

Brain Scan #2. *Ketamine enhances structural plasticity in human dopaminergic neurons: possible relevance for treatment-resistant depression.* Collo G, Merlo Pich E (2018) *Neural Regeneration Research 13(4):645-646.* www.nrronline.org

> Defective functional and structural plasticity in the glutamatergic frontocortical and hippocampal circuits has been associated to high levels of circulating glucocorticoids and to reduced central levels of brain derived neurotrophic factor (BDNF) (Autry et al., 2011; Duman et al., 2016). Intriguingly, downregulation of intracellular molecular pathways involved in cell growth and survival, *i.e.*, the mammalian target of rapamycin (mTOR) pathway, was described in postmortem prefrontal cortex of patients with mood disorders (Jernigan et al., 2011). Defective neuronal structural plasticity

was also observed in frontocortical and hippocampal circuits of rodents after chronic stress, phenomenon reversed by repeated electroconvulsive therapy, chronic SSRI treatments or, as shown more recently, by acute infusion of ketamine...

These results suggest that the prolonged antidepressant effects observed after a single infusion of ketamine in Treatment Resistant Depression (TRD) patients can be related to a transient enhancement of structural plasticity induced by a glutamate "burst" occurring not only in frontal and hippocampal neurons but also, in mesencephalic DA neurons. The known interconnectivity between these 2 circuits suggests a potential crosstalk. Since Major Depressive Disorder and TRD have been associated to defective neural plasticity, the structural plasticity induced by these treatments could be interpreted as a potential remediation of an underlying neurobiological mechanism that sustains depressive symptoms.

Motivation to Change: *From the Sense-of-Self to the Global Village*

During the Covid-19 pandemic I published *Global Village Burnout: Creative Intelligence = Survival* (2021). The book reflects on the escalating patterns of social self-sabotage. Human beings have profound capacity for changing themselves. However, they require considerable motivation to make a sustained effort. The chief incentive to change is fear of loss.

As we expand our knowledge base of how stress triggers self-sabotage in individuals, families, and society, perhaps we can choose to support motivation for change to prevent catastrophic events like war, pandemic, and racial violence. We know that the attainment of significant numbers of people to kindle such a cultural evolution must begin in early childhood in families and education. It will take generations to evolve a collective sense of *Species Tribalism*, the biological fact that human beings are a single race. Species Tribalism remains an obscure social concept.

RAZOR'S EDGE

I live at the razors edge,

The tipping point,

Between the Newtonian cosmos and the

Quantum realm.

These is no time or space,

No life or death,

No past or future.

There is only now,

And now is constantly changing.

Balancing the mind-body

Empowers my joy of living,

Expands my love of all life,

And deepens my gratitude for the

Nature of my existence.

NOH TIME #3 (Front)

1992, Japanese lacquer on wood, bamboo, steel rings, washi, gold leaf, 26" x 28"
Phillip Romero (Photographed by Jennifer Douglas)

NOH TIME #3 (Front)

1992, Japanese lacquer on wood, bamboo, steel rings, washi, gold leaf, 26" x 28"
Phillip Romero (Photographed by Jennifer Douglas)

I

<u>The Logosoma Living Wellness Protocol</u>

Marilyn Monroe at home, 1953, Alfred Eisenstaedt, Life Pictures, Shutterstock

"Creativity has got to start with humanity and when you're a human being, you feel, you suffer."

Marilyn Monroe

Logosoma Living Wellness Protocol

The *Logosoma Living Wellness Protocol* (LLWP) is a guide for mindful integration of the story of our lives and the body we inhabit—*Logosoma*—the abstract *logos* (word, story, rational principle) and the concrete *soma* (body). Before we are conceived, the story of us lives in the imaginations of our parents. After our body ceases to be, the story we lived will be remembered by those that knew us. We can see that the body follows the story, and the story persists after the body is gone.

Experience teaches us that we live at a tipping point between positive, life supporting events and negative life-challenges—creativity against adversity. Logosoma Living provides skills to navigate negative stress for balanced living through mindful awareness of the story we create and loving kindness and gratitude for the gift of our body. The meaning we create for ourselves and for others emerges from an authentic narrative of seamless integration of the body and the story.

Where did Logosoma Theory come from?

As a practicing family psychiatrist in New York City with cross-cultural interests, I understood how stress triggers breakdowns in family systems. I developed an empirical theory to answer the question, "What is a human being?" My answer was, "A human being is a body living in a life story told by the body that has profound effects on the body." My training protocol, *Ways of Listening Toward an End of Suffering,* derived from Buddhism's Four Noble Truths and focused on stress regulation in individuals and families. I also used the techniques to train Family Psychiatry Fellows as Assistant Professor of Child Psychiatry at New York-Presbyterian Hospital, Weill Medical College-Cornell University. Integrating Buddhism with developmental psychobiology and systems theory, I needed a proprietary name for the protocol.

I had been pondering the creation of a name by looking at Latin and Greek root words: soma, lingua, corpus, oratio, logos, etc. In a dream, I found myself back in med school and being quizzed on the word for body and story. After fumbling for words, I came up with *'Logosoma'*. The professor said, 'right', and I woke up with the name for my technique and the beginning of Logosoma Theory. Later in studying the neuroscience of creativity, I would come to know what I had always felt as an artist since childhood, the brain is a creativity machine. Logosoma Theory would be grounded in neuroscience and mindfulness practices to mitigate relationship attachment stress, the root of human suffering, and promote the development of an authentic sense-of-self.

Cross-Cultural Medicine: *My Story*

My pre-medical education was a survey of the critical domains of knowledge required to study medicine: biology, anatomy, biochemistry, physics, psychology, and any electives one chooses. Medical school delves deeply into these areas, expanding into the physiology of life, the cellular dynamics, organs, and organ systems, and how life teeters between health and pathology. Research drives the ever-increasing technologies that provide a scientific foundation to advance therapeutics in medicine.

My own research led me to comparative cultural medicine. The University of Texas Medical Branch at Galveston had an extensive History of Medicine Library and a department of Medical History and Philosophy, headed by Emeritus Professor William Bean, MD. They offered the Chauncey Leake essay contest, named after the institutions founder. I won the prize with my essay, *Holistic Medicine*. I reviewed the role of Paracelsus, an ex-communicated alchemist who was among the founders of allopathic medicine with his discovery of the mercury treatment for syphilis. Jungian psychology framed alchemy as a symbolic transformation of the sense-of-self. Paracelsus predicted the emergence of allopathic medicine and the Age of Enlightenment.

Dr. Bean encouraged my research the following summer when I decided to go to India for my three-month externship to study at the Yoga Institute in Bombay, and Tibetan Medical Center in Dharamsala. At the Yoga Institute, I studied Patanjali's sutras with an introduction to Ayurvedic medicine. During my studies in Tibetan Buddhism, meditation, medicine, and culture, the HH Dalai Lama gave me letters of introduction for further teachings with Thrangu Rinpoche and Tenga Rinpoche in Kathmandu, Nepal. My essay, *Tibetan Medicine: An Ancient System Struggles to Survive*, won the contest. I was offered an externship at Pan Am University in Edinburgh, Texas, to study Curanderismo, a synthesis of spiritual and religious disciplines that form indigenous healing practices. I worked with a Professor of Anthropology, shadowed Poncho, a Curandero, as he held a clinic and did home visits in the barrios. I wrote another essay.

During my Residency in psychiatry at St. Vincent's Hospital, New York, I invited my teacher, Tenga Rinpoche, to give a Grand Rounds conference on Tibetan Buddhist psychology. I also presented my research on Tibetan Medicine at the Jacques Marchais Museum of Tibetan Art in Staten Island, New York City.

My career as an author was kindled by my perpetual curiosity about bridging domains of knowledge, especially in the links between art-mind-culture topics, artists, and scientists. Many years later, when E.O. Wilson wrote *Consilience: The Unity of Knowledge* (1998), I discovered one of my passions was consilience. I asked Ed Wilson to review my work on *The Art Imperative: The Secret Power of Art* (2010). Ed is Pellegrino University Research Professor at Harvard, author of two Pulitzer Prize winning books, *On Human Nature* (1977) *The Ants* (1990). Ed wrote a cover comment for my book: *"This attractive and interesting work is an important venture into the borderland between the creative arts and science."*

The Tao of Wellness: *Consciousness Beyond Pathologizing*

Logosoma Training integrates diverse models of wellness as a mindfulness practice. Both holistic and allopathic medicine offer frameworks that contribute to a biopsychosocial model for human wellness. All systems place stress-regulation at the center of wellness. The allostatic system that regulates stress response is critical to survival and reproduction. Life is a complex system of perpetual change striving for a dynamic balance.

Allopathic medicine, a system in which medical doctors and other healthcare professionals (such as nurses, pharmacists, and therapists) treat symptoms and diseases using drugs, radiation, or surgery, emerged during the Enlightenment—the 'Age of Reason.' This period of rigorous scientific, political, and philosophical discourse characterized European society during the 18th century. Fact-based theories drive scientific and mathematical exploration that produce technological advances allowing us to create improved living conditions for humanity and explore the known universe.

Holistic medicine is prehistoric, and ancient healers used complex systems theories and beliefs to understand and treat the maladies of the mind and body. These systems have not been adequately subjected to scientific exploration until recently. Logosoma theory strives toward balanced integration of these approaches.

Consciousness: *The Final Frontier*

The neuroscience of consciousness is generating a paradigm shift about wellness. The scientific romance with 'facts' is being displaced with new, complex systems models that explore our perceived reality. What we feel and think has more to do with wellness than we realize. Patterns of negative thinking emerge in insecure attachments, and they can have profound effects on undermining wellness at the molecular level.

Ancient models of wellness rely heavily on narratives of hope, belief, and faith. These concepts are aligning with new understandings of how the brain perceives and constructs personal reality and creates meaning in the universe. This narrative can have positive benefits on stress regulation and detoxification of 'burnout' physiology—the result of toxic, chronic stress.

Many neuroscientists see consciousness as a by-product of the neural networks of the brain—*emergentism*. Mindfulness studies from ancient contemplative traditions, Taoism, Buddhism, Platonism, produced philosophic paradigms that describes consciousness from a holistic viewpoint, *panpsychism*. Consciousness is seen as a universal phenomenon that imbues all self-organizing systems.

Nobel physicist Roger Penrose and anesthesiologist Stuart Hameroff propose the Orchestrated Objective Reduction theory of consciousness (Orch OR). They speculate that consciousness emerges from 'quantum collapses' occurring in the microtubules in the cytoskeletal structures of living cells. This consilient theory, if proven, would reframe our brain centered view of consciousness by integrating emergentism and panpsychism. It places consciousness at the interface between the two domains of the known universe, the classic domain described by Newtonian physics and the quantum domain predicted by Max Plank, Albert Einstein, and others.

The more we understand the mechanism of consciousness, the greater our skills in mindful self-regulation. Our emotions and learned beliefs, consciously and non-consciously, drive many self-destructive behaviors. Relationship stress is the cause of family dysfunction from domestic violence to divorce. Socioeconomic disparity stress feeds the social structures of greed, racism, sexism, and religious intolerance. The last 10,000 years of human civilization reveal how profoundly creative we can be. They also illuminate how emotionally ignorant we are and how self-destructive we can be to each other and to the biosphere.

Three Modules of *Logosoma Living Wellness Protocol*

In Part II, the first module, *Logosoma Brain Training* (LBT), focuses on mindful self-regulation of relationship stress. *Phantom Stress: Brain Training to Master Relationship Stress* (2010) presents LBT is detail. The 4-Rs algorithm helps the brain-mind rewire the stress response to many stress triggers: *Refuse, Refocus, Reflect, Reconnect.*

LBT is based on attachment theory. Ancient Buddhist philosophy, the first attachment theory, is integrated with modern attachment theory, proposed by John Bowlby and others. Together, they form the developmental psychobiology and social neuroscience basis of LBT. The work of neuroscientist Michael Meaney reveals the role secure attachment plays in early development that shapes the molecular basis of the stress regulatory system across the lifespan. Insecure attachment can be rewired in adulthood due to the brain's neuroplasticity. Part II reviews and explains the 4Rs algorithm.

Part III explores the second module, *Creative Intelligence Training* (CIT). CIT emerged from LBT with the 6Rs: *Remember, Reflect, Reframe, Reimagine, Reinvent, Reconnect.* After the Covid pandemic, many of my former patients returned with symptoms of 'burnout' in careers and families. I introduced CIT in *Global Village Burnout: Creative Intelligence = Survival* (2021), written during the Covid pandemic.

The practice of CIT helps remodel the neural networks of the sense-of-self. The use of guided meditation, daydreaming, philosophical contemplation, and creative practices from art to music to writing, engage the brain's neuroplasticity that contributes to renewal of the sense-of-self. These practices train the mindful awareness beyond the sense-of-self, what I call *selfless mind.* Ancient mindfulness practices focus on attaining selfless mind as the way to 'end suffering.' Transcending the self-centered ego by cultivating selfless mind mitigates the anxiety of impermanence. The fact that each individual lives in their own private reality, 'relative reality,' triggers confusion and relationship stress. CIT trains empathy, emotional intelligence, and supports the transformation from anxious 'ME' to secure attachment in 'WE'.

In Part IV, the final module, *Psychopoiesis Training* (PT) integrates newly available psychedelic medication, ketamine, into Logosoma Training. The potential goals of psychedelic medicine empower mastering the *anxiety of impermanence* and rewiring the *sense-of-self*. Ketamine pharmacologically initiates a rapid induction of the experience of selfless mind by disrupting the default mode network (DMN), the 'resting sense-of-self.' This activates a window of heightened neuroplasticity for learning. Both implicit and explicit learning can be empowered with PT during the ketamine treatment. Psychopoiesis training creates a ramp of mindfulness learning before, during, and after the ketamine experience.

NOH TIME: HANDS #1 (After Michelangelo)

The ancient hand,

Charged with invisible forces,

Swirls the cosmic dust

Out of darkness into

Luminescence,

Spiral galaxies, gaseous frescos, and

Exploding points of sacred light.

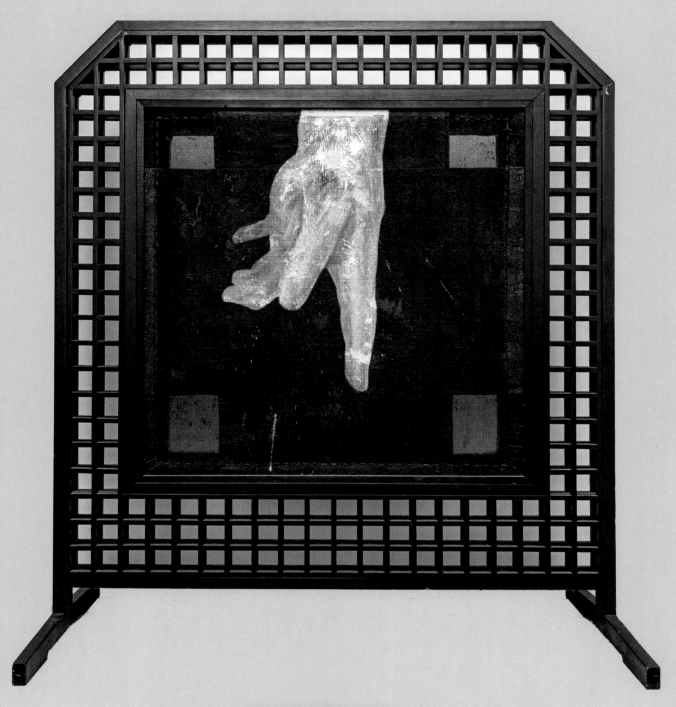

NOH TIME #2 (Back)

(After Michelangelo), 1988, 43" x 49," Wood, Oil/canvas,
Phillip Romero (Photographed by Jennifer Douglas)

NOH TIME #2 (Back)

(After Michelangelo), 1988, 43" x 49," Wood, Oil/canvas,
Phillip Romero (Photographed by Jennifer Douglas)

II

<u>LOGOSOMA BRAIN TRAINING</u>

The Body and the Life Story

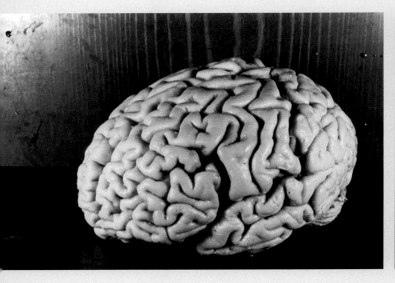 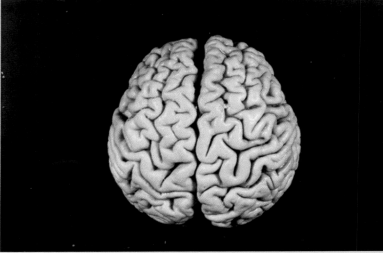

"Left lateral view of Einstein brain, prior to sectioning," Otis Historical Archives, *National Museum of Health and Medicine.*

"Superior view of Einstein brain, prior to sectioning," Otis Historical Archives, *National Museum of Health and Medicine.*

"Imagination is more important than knowledge. Knowledge is limited. Imagination encircles the world."

Albert Einstein

Logosoma Brain Training (LBT) integrates five mindfulness exercises to recruit the brains neuroplasticity for changing the sense-of-self in relationships. The Identity Declaration (2.1) is an empirical guided meditation. The Mindfulness Script (2.2) is a guided meditation that distinguished the Emotional Hostage Burnout Script of the Limbic system brain structures from the mindful narrative of self-creation guided by the prefrontal cortex (PFC). The 4Rs to Master Relationship Stress (2.3) guides new patterns of thinking and mindful reflection to navigate the self-sabotaging emotional reactions to relationship stress. The VED (2.4) presents a transactional mindfulness framework for mastering negative assessments in relationships. The Relationship Maintenance Conversation (RMC) (2.5) is a template to guide the detoxification of stress in relationships. With practice, the RMC prevents codependency and burnout.

2.1 IDENTITY DECLARATION

Each human being develops a conscious sense-of-self that is experienced as "me." Grounded in the body, this awareness of self is first and foremost an emotional self. Brain-mind development takes about 25 years to complete the growth of neural networks for high level cognitive functioning. Sexual and reproductive maturity occur in mid adolescence. The prolonged dependence on parents, families, and culture for the brain maturation of the individual identity creates many vulnerabilities in adult attachment to one's body and to other selves.

Mindful awareness of this process can promote acceptance of the emotional stressors in attachment. To mitigate the inevitable problems in identity development and promote healthy relationships, the *Identity Declaration* exercise was developed as fundamental part of Logosoma Brain Training. Written with a phenomenological approach, the declaration is designed to be read slowly out loud. Listening to oneself utter the words kindles memories and associations for reflection. Exploring the feelings and thoughts that emerge activates mind-body awareness in the present moment. Readers can record their voice and use it as a guided meditation.

I

What am I?

I, *my name*, am a human being.

I live in an infinitely complex body

endowed with unique characteristics,

and I also live in a life story.

My brain is a creativity machine.

It spontaneously creates images,

Emotions, ideas, memories, and stories

about my experience.

My sense of 'self' emerges from my attachments

To people, places, and things

from birth and across my lifespan.

I instinctively strive toward secure attachments to my body, to my

relationships, to my community, to my job and career activities, to

my culture, and to my environment.

With inborn *Creative Intelligence,* I make choices

that shape my sense of identity and

drive the story of my life.

II

Who am I?

My life story is created in two distinct narratives.

I live in a concrete, emotional story, grounded in my body's

experiences, and

I live in an abstract, cognitive story in which

I use my memory of the past and my imagination

about the present and the future to create meaningful

connections that help me make sense of being Me.

These two narratives are perpetually intertwined consciously

and non-consciously across my lifespan.

My emotional story springs from my body.

My body, with its five senses, experiences the present, moment

by moment.

I can experience a vast spectrum of emotions from pain to pleasure.

My emotions shape my thinking, my attitudes, and my behaviors.

With conscious awareness, what I call my *mind,* I can pay attention

to my emotions.

III

My Mind-Body connection

Using language, I can name the emotional states of my body

with words called 'feelings.'

My awareness of feelings helps me make sense of what my body

is experiencing.

With practice reflecting on my emotions, I am better able to

create a story of Me, of how I feel and what I think about my life

story.

With practice, I can improve my skills of being

mindfully aware of myself, of my body and my story.

My self-awareness empowers me to create an authentic sense

of my identity.

With mindful self-awareness, I can regulate my emotions,

and guide myself in creating secure attachments across my life.

Practicing mindfulness opens my

experience to the joys of being alive,

to the wonderment of perception, and

to the richness of my imagination's creative possibilities.

IV

Life is stressful.

I accept that life is stressful.

Stress can be positive and negative.

My body has a sophisticated stress regulatory system that helps me survive the negative stresses and enjoy the positive stresses.

I am grateful to have a brain and mind that can learn to master the stresses of life.

I can overcome life stresses with my natural creativity.

My brain is a creativity machine, and with practice, I empower my creative nature with my mind.

From the day I am born my creative brain begins to explore the world, practice learning to move my body, experiment with crawling, walking, talking, and learning to play.

V

Positive Stress

Positive, excitement stress, inspires my creativity.

With my creativity I can play, from childhood and beyond.

Play activates my body's action and my mind's imagination.

With curiosity I explore myself, other people, places, and

things. I can invent new ways of doing things in play.

I can experiment with my life, with the world, and

with relationships.

Taking risks inspires excitement and learning.

When I practice play, my skills get better, and my

competence improves.

Playing alone nurtures my imagination.

Playing with others by taking turns and sharing ideas and

actions empowers and strengthens trust and serious fun.

VI

Negative Stress

Negative stress threatens my body and all my attachments.

My body is wired to react to perceived threats with emotions that move my body to freeze, fight, or flee the threat.

I can feel threatened in my attachment to my body, in my emotional relationships with others, and in my thoughts about the past, the present, and future.

Negative stress can paralyze my creativity.

It can be episodic, and it can become chronic.

When stress is episodic, it can teach me how to react when it occurs again.

I learn resilience from episodes of stress, and it empowers my skills for competent living.

When stress becomes chronic, it can wear down my brain's creativity.

With perpetual stress, I can feel like a hostage.

My optimistic attitudes can become impaired, and my behaviors can become self-sabotaging.

Chronic stress can become toxic to my body, and my brain can be hijacked by stress hormones.

Toxic stress leads to physical, emotional, and cognitive burnout. My ability to function in the world breaks down.

VII

Creative Intelligence

Creative Intelligence is a brain-mind practice.

With mindful coordination of

My inborn resilience to adversity and the

Creativity networks of my brain,

Creative Intelligence empowers resilience to toxic stress.

With focused attention, I can learn to regulate my mind-body

perception of stress and activate my creativity.

By reframing my perception of toxic stress,

I can redirect my emotional reactions in my attachment to others.

And I can overcome the 'phantom stress' of past experiences.

I can learn to master my fear of future stress.

With Creative Intelligence

I activate my imagination to project possible futures

Where my renewed sense-of-self rediscovers

The wonders of being and doing.

VIII

Self-making is a wish-driven process

Creative Intelligence kindles the first step in creating a new story of my Self.

My sense-of-self is an ever-evolving process.

Self-making begins in infancy and continues across my life with each new memory.

I create memories from observing the world, from reacting to the world, and from exploring the world with my curiosity.

As a body living in a life story, I can create relationships with others that change my perception of reality and expand my sense of ME into the fabric of WE.

The meaningful connections I create with people, places, and things empower my creativity to create a true sense of my purpose in life.

2.2 MINDFULNESS SCRIPT

The Mindfulness Script template provides an exercise for distinguishing the emotional brain script (the right-side column) from the mindful cognitive mindful script (left-side column). The limbic system, the emotional brain, forms the ancient mammalian survival system: freeze, fight, flight during times of danger, and when safety is perceived, forage for food to survive and mate reproduce—the *5Fs, freeze-fight-flight-feed-fornicate.*

When the limbic brain is chronically stressed, *The Hostage-Burnout Script* emerges. Even a single traumatic episode can flood the limbic brain with the stress hormone, adrenalin, damaging the ability to recover. When the limbic brain is triggered repeatedly, it produces toxic levels of cortisol and an overactive inflammatory response in the immune system. Posttraumatic stress and toxic stress eventually take a toll on total body function producing the well-known 'burnout syndrome'. Symptoms of burnout include hypervigilance, irritability, short-temper, anxiety, insomnia, flashbacks, depression, self-sabotaging addictive behaviors, and relationship dysfunction.

LBT recruits the Mindfulness Script to refocus and reframe the instinctive stress. LBT down-regulates the freeze-fight-flight response with the relaxation response to calm-connect-create. Mediated by mindfulness practices, oxytocin triggers a cascade of neuroendocrine mind-body reconnections.

The Mindfulness Script Exercise

Begin with a deep breath, as when you begin mediation. Focus your attention on the right column, the Burnout Script. Read each box aloud, pausing to reflect on how it feels to utter the words. Let you association to memories when you felt this way. Remember, you are giving your limbic brain speech. Your body is speaking. You are listening to the body declare its perceptions. After each reflection, read the next box aloud and repeat the reflections on feelings and thoughts about what your body is telling you. After finishing the entire script pause and meditate for a while with loving kindness toward your body. Your body, like a child, just reported the experience of that cascade from stress trigger to burnout.

Now refocus on the Mindfulness Script. Repeat the exercise of breathing, reading aloud, reflection on your feelings and thoughts, allowing associations to wander. Slowly work through the eight steps of the mindfulness script. Reflecting from the abstract mind with compassion and gratitude for your body.

Finally compare the highlighted words in the right and left columns. These are the dialectics of the mind-body narratives. They always exists as potential opposites. You cannot hold them both in mind at the same time. Your ability to reframe the language shuts OFF the stress response circuit and trips the relaxation response switch ON.

When you become aware of 'what IF' thinking, you can mindfully reframe to 'what IS' to short-circuit the stress reaction cascade and calm your body with reassurance from the compassionate mind. The more you practice the mindful script, the easier it becomes.

Eventually you can rewire you stress trigger rection toward calm, delayed, mindful assessments, rather than acting out the freeze-fight-flight physiology and behavior. Research shows that mindfulness meditation increases the time to focus on behavioral choices after stress triggers. The anatomic path from the limbic brain to the mindful brain becomes much easier to access with practice.

Practicing the Mindfulness Script will change your brain's reaction to stress and empower expanded awareness of your body.

MINDFULNES SCRIPT (Conscious, abstract, reflective attention)	HOSTAGE BRAIN-BURNOUT SCRIPT (Non-conscious, concrete, emotional physiology)
"Sometimes I *feel like a* Victim, and I am able to mindfully reframe the experience."	"My body is telling me I'm not safe, and I feel frightened, angry & confused:"
1. I am **Not a Victim**.	1. I am a **Victim**.
2. I **Accept** this experience is difficult, and I am **Grateful** that I can reframe my attitudes and behaviors in response to the situation.	2. I feel **Self-pity** and **Resentment** for the person/ circumstance that made me feel this way.
3. I **Refocus** my attention to **What Is** triggering my body in the present and slow my breathing to calm myself.	3. I can only think of **What IF** something worse happens next.
4. I can **Manage myself** in the situation to the best of my abilities.	4. I need to **Control** the situation and/or others.
5. I **Wish to** create a new narrative connection with the people, places, things that triggered my stress.	5. I **should do**…or they should do…to make the situation right, because it is wrong now.
6. I want to **Practice #1-5** to **gain competence** in mindful self-regulation when I'm triggered to feel like a victim.	6. I need to keep **Trying** to **Succeed**…I can't **fail**.
7. Sometimes, even when I do my best, my brain gets hijacked and I will act out my anger, fear, and confusion… **"It's not ME it's my brain on stress."**	7. No matter how hard I try, I keep **Failing**…
8. Even in the worst of times, I accept that **all things are Impermanent**. These things, the good, the bad, the sad and happy times, will pass.	8. I feel burnt out…I feel **Resignation**.

2.3 4RS TO MASTER RELATIONSHIP STRESS

The 4Rs to Master Relationship Stress is a mindfulness clinical practice created in 1986. For over thirty years I have taught the algorithm to help people rewire their brain's stress response and create new neural networks for secure attachment. Practicing the 4Rs empowers multiple changes in your relationship experience. They include:

- Identify stress triggers earlier
- Change your reaction to the trigger
- Refocus your attention for mindful awareness of your body
- Reflect on the emotions of stress: freeze-fight-flee
- Name the physical emotions with words: confusion, anger, fear
- Turn OFF the stress response
- Turn ON the Relaxation response
- Integrate your authentic mind-body self
- Explore selfless mind
- Reconnect with your partner
- Develop the skills for reciprocal validation and empathy
- Discuss the issues that trigger stress to detoxify resentment
- Prevent toxic stress and codependency

Using the 4Rs Algorithm

The 4Rs algorithm is best used as a study guide. Learning it alone and discussing it with others helps imbed the patterns in long-term memory. Once you have learned the steps, the details of 'how to' implement it can be practiced with your partner and family.

The 4Rs, *Refuse, Refocus, Reflect, Reconnect,* are easy to remember. Each word guides your mindful attention to soothe your body's stress response, to take an emotional inventory, and to eventually reconnect with your partner. Empowering the Rs with simple wish-driven statements makes them robust mindfulness declarations.

1. I *Refuse* to let my stress hormones hijack me.
2. I *Refocus* my attention to soothe my body.
3. I *Reflect* on my anger, fear, confusion, and ask myself, "What do I truly want?"
4. I *Reconnect* with my partner by *validating* their experience and *empathizing* with their perceptions.

Learning each step of the algorithm will transform the idealized wish-drive statement into a practical skill to immunize your relationship to toxic stress.

Logosoma Brain Training
Rewiring Your Brain's Stress Response

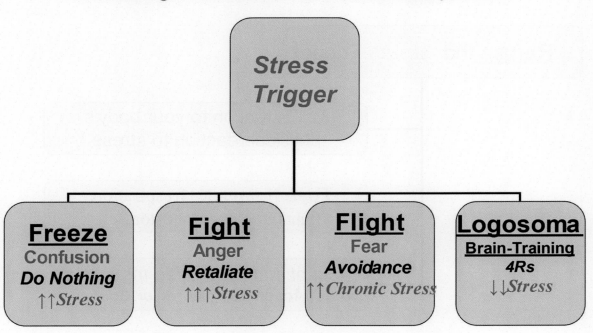

STEP 1: REFUSE

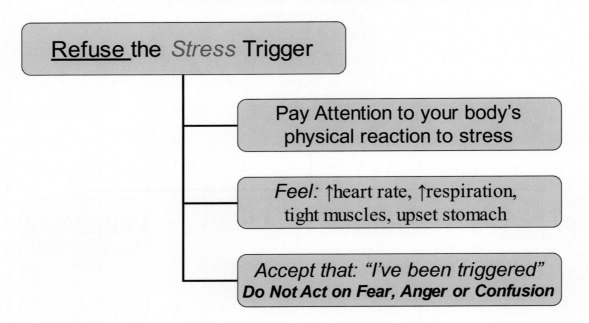

Refuse the *Stress* Trigger

Pay Attention to your body's physical reaction to stress

Feel: ↑heart rate, ↑respiration, tight muscles, upset stomach

Accept that: "I've been triggered"
Do Not Act on Fear, Anger or Confusion

STEP 2: REFOCUS

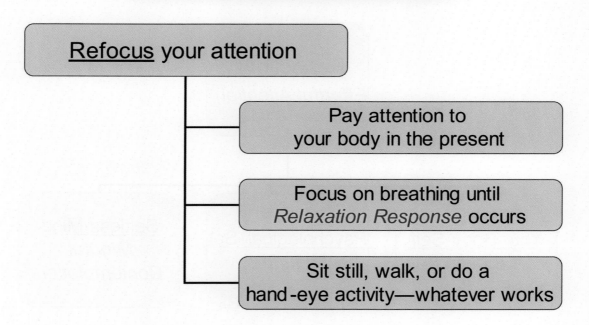

Refocus your attention

Pay attention to
your body in the present

Focus on breathing until
Relaxation Response occurs

Sit still, walk, or do a
hand-eye activity—whatever works

STEP 3: REFLECT

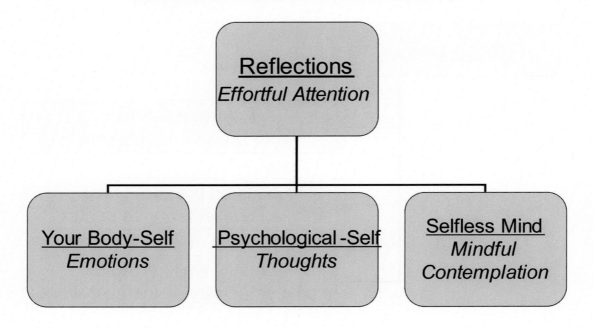

STEP 3-A: Your Body-Self

After calming your body, turn your attention inward to your *Emotions*

Reflect on your *Fear*
"What am I afraid of?"

Reflect on your *Confusion*
"What triggered my stress reaction?"

Reflect on your *Anger*
" What am I angry about?"

STEP 3-B: Your Psychological Self

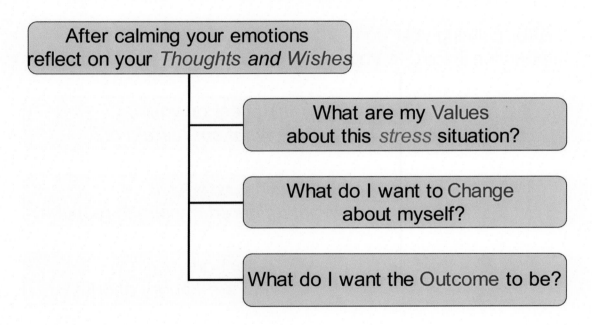

After calming your emotions reflect on your *Thoughts and Wishes*

What are my Values about this *stress* situation?

What do I want to Change about myself?

What do I want the Outcome to be?

STEP3-C: Your Psychological Self

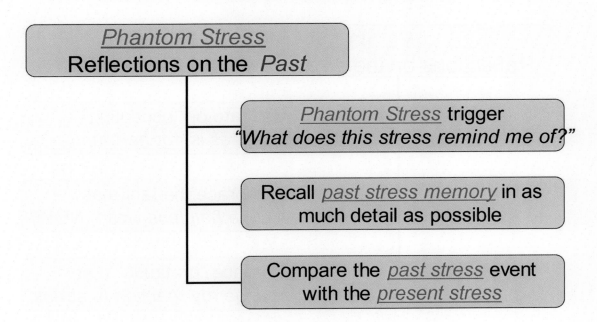

Phantom Stress
Reflections on the *Past*

Phantom Stress trigger
"What does this stress remind me of?"

Recall *past stress memory* in as
much detail as possible

Compare the *past stress* event
with the *present stress*

STEP3-D: Your Psychological Self

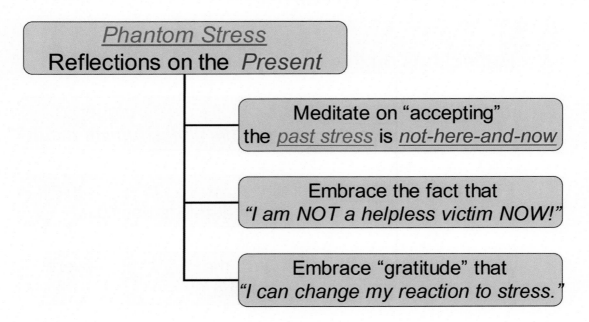

Phantom Stress
Reflections on the *Present*

Meditate on "accepting"
the *past stress* is *not-here-and-now*

Embrace the fact that
"I am NOT a helpless victim NOW!"

Embrace "gratitude" that
"I can change my reaction to stress."

STEP 3-E: Selfless Mind

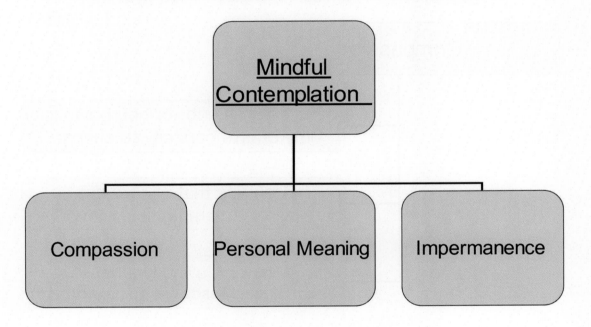

STEP 3-F: Selfless Mind

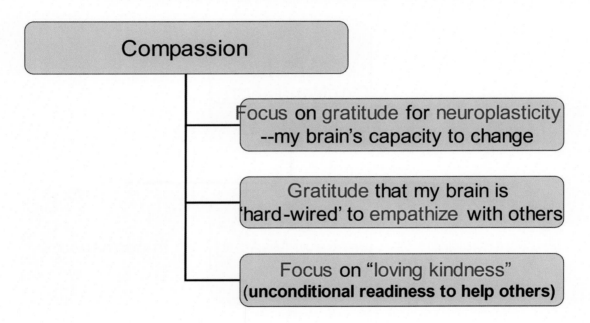

Compassion

Focus on gratitude for neuroplasticity --my brain's capacity to change

Gratitude that my brain is 'hard-wired' to empathize with others

Focus on "loving kindness" **(unconditional readiness to help others)**

STEP 3-G: Selfless Mind

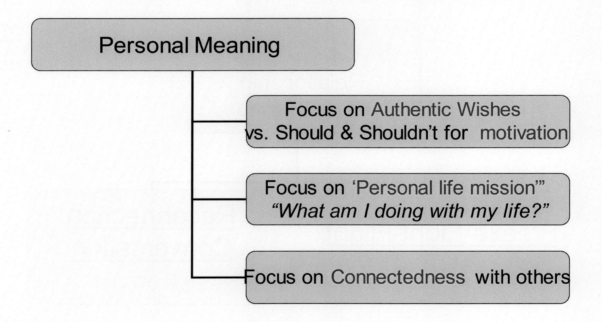

Personal Meaning

Focus on Authentic Wishes
vs. Should & Shouldn't for motivation

Focus on 'Personal life mission'"
"What am I doing with my life?"

Focus on Connectedness with others

STEP 4: RECONNECT

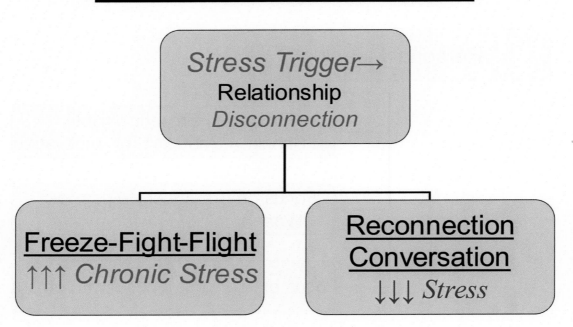

STEP 4-A: Reconnection Conversation

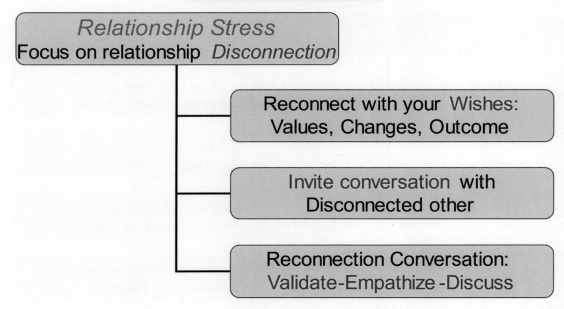

Relationship Stress
Focus on relationship *Disconnection*

Reconnect with your Wishes: Values, Changes, Outcome

Invite conversation with Disconnected other

Reconnection Conversation: Validate-Empathize-Discuss

STEP 4-B: Reconnection Conversation

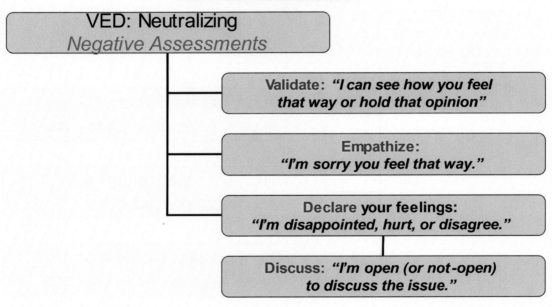

VED: Neutralizing
Negative Assessments

Validate: *"I can see how you feel that way or hold that opinion"*

Empathize: *"I'm sorry you feel that way."*

Declare your feelings: *"I'm disappointed, hurt, or disagree."*

Discuss: *"I'm open (or not-open) to discuss the issue."*

STEP 4-C: Reconnection Conversation

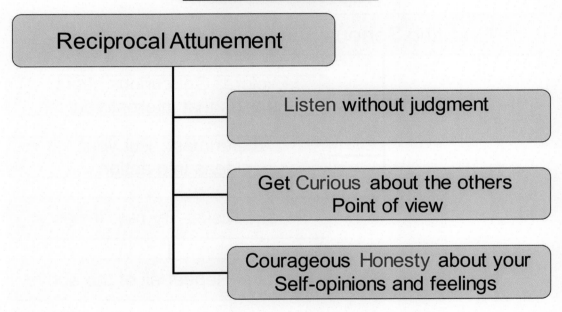

Reciprocal Attunement

Listen without judgment

Get Curious about the others Point of view

Courageous Honesty about your Self-opinions and feelings

STEP 4-D: Reconnection Conversation

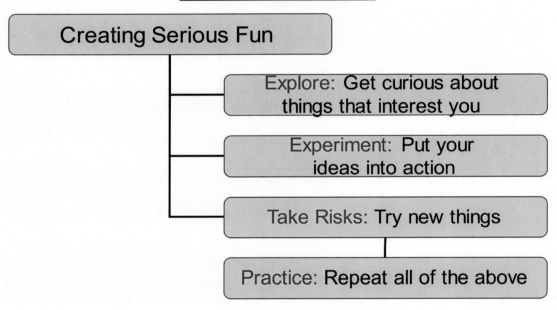

Creating Serious Fun

Explore: Get curious about things that interest you

Experiment: Put your ideas into action

Take Risks: Try new things

Practice: Repeat all of the above

STEP 4-E: Reconnection Conversation

Creating Positive Memories

Revisit positive experiences in conversation with others

Remember positive experiences alone and with others

Archive positive experiences with photos, journals, memorabilia

2.4 VED: *Validate, Empathize, Declare-Discuss*

Neutralizing negative emotions in relationships is critical to a sustainable system of two or more members of a family. The VED is a mindfulness skill for immunizing yourself to the emotions triggered by *negative assessments*. Based on self-respect, the VED empowers *self-regulation*. The VED helps you master the *freeze-fight-flight* response to criticism and blame.

To navigate negative emotions in a relationship, you must identify what triggers stress in your body—namely *negative assessments*. Negative assessments can be either *explicit* or *implicit*. Examples of explicit negative assessments include derogatory attacks on your appearance, behavior, or beliefs. Implicit negative assessments include gestures, diverting attention away, eye-rolling, and shrugging. Both trigger the freeze-fight-flee physiology and hijack your brain.

Negative assessments trigger feelings of shame, and prompt defensive and retaliation behaviors. They often trigger *Phantom Stress* through nonconscious associations to childhood memories. Phantom stress amplifies the present stress to inappropriate levels and triggers a cascade of escalating relationship disconnections. This interaction is common, and it can devolve into patterns of codependency if not checked.

Remember, if you feel like a hostage, your partner probably does, too.

Living in the VED: *Toward the Reconnection Conversation*

Cultivating the practice of reciprocal mindfulness with the VED promotes detoxifying relationship stress and prevents codependency.

When you agree to this guideline, you are ready to communicate. The VED can be used spontaneously, and it can be scheduled as a maintenance practice. The VED Script:

- **Validate:** *"I can see how you felt criticized by what I said."*

Validation of your partner's feeling is a self-respecting beginning. Without it both partners will simply resume playing the victim/oppressor scenario and the verbal warfare will resume.

- **Empathize:** *"I'm sorry you feel this way."*

Empathy is emotional mirroring, the ability to feel someone else's pain. If you can't empathize with your significant other, what's the point in continuing this discussion? When we feel empathy and our pain is validated, we can down-regulate into a reflective mind. Now we are ready to reconnect.

- **Declare &/or Discuss:** *"Is there anything I can do to help you?"*

By declaring your willingness to help, you identify yourself as an advocate, not an adversary. Being open to discuss an honest assessment of what really set you off, you share your feelings. This declaring and discussing function as an invitation to your partner to do the same. Reciprocal empathy and validation mirror your separate identities. This technique that recruits compassion for the We-team.

Living in the VED trains your brain to be open, non-judgmental, and empathic to others. When you and your partner refocus together, you trigger a cascade of positive bonds that include:

- openness to accept responsibility for hurting each other

- showing a willingness to make amends
- paying reflective attention to what occurred
- practicing reassurance
- active listening in a turn-taking process with no losers
- increasing the neural networks of patience and compassion

Now you can resume the path of making a "WE" story together.

2.5 The Relationship Maintenance Conversation

The Relationship Maintenance Conversation (RMC) is a practice in joint attention and reciprocal communication. The RMC provides your We-Team with a skill to detoxify *negative emotions*, with a focus on *Fear*. Fear is a feeling triggered by stress and is always accompanied by *anger* and *confusion*. These negative emotions can trigger disconnections in communication. They occur naturally when disappointments in expectations occur. When We-team stress becomes chronic, the pattern of codependency emerges with resentment, self-pity, and blame. Codependency undermines trust and respect in human relationships. The primary goal of the RMC is creating habits for *better listening.* Consider this the planning room for strategic solutions for the stresses of everyday life and phantom stress. Remember, stress, not your partner, is your enemy. The We-team is an alliance against the toxic stress that can emerge naturally in families. Like weeds in a flower garden, regular stress detoxification is optimal. The RMC format provides a guide for turn-taking conversations. By using a turn-taking structure for these meetings, you agree beforehand to inhibit your urge to interrupt the speaker. Focus on listening to your partner. Mindful effort to apply your VED training to listening is critical. You will have a chance to reflect on your reactions to their comments later. Let your partner finish what he or she is saying.

Example

Bob reports two fears about himself, his workaholic schedule, and his avoidance of emotional conversations. He shares one fear about Alice, the fear that she will leave him. His fear about them as a couple is their children will suffer.

I instruct Alice to listen silently until Bob is finished. Taking notes often helps the listener to focus. Then I ask Alice to respond with her feelings, observations, and questions. Bob must listen until she is finished. The turn taking method develops empathy and patience. The meeting concludes when both of their issues have been discussed to their mutual satisfaction, or until they agree to continue the meeting at another time. Each meeting should end with rescheduling the next meeting.

The Conversation Template Each set of boxes (fears, concerns, requests) is considered one conversation topic. A 10–15-minute time limit is recommended for each set of boxes.

Mine	Yours
Fears What feelings of *fear* do you have about **yourself** in this partnership? What feelings of *fear* do you have about **your partner**? What feelings of *fear* do you have about **your partnership**?	**Fears** What feelings of *fear* do you have about **yourself** in this partnership? What feelings of *fear* do you have about **your partner**? What feelings of *fear* do you have about **your partnership**?
Concerns What **observations** concern you about **yourself** in your partnership? What observations concern you about your **partner**? What observations concern you about your **partnership**?	**Concerns** What **observations** concern you about **yourself** in your partnership? What observations concern you about your **partner**? What observations concern you about your **partnership**?
Requests What requests would you like to make to your partner about **their self**? What requests would you like to make about the **partnership**?	**Requests** What requests would you like to make to your partner about **their self**? What requests would you like to make about the **partnership**?

NOH TIME: HANDS #2 (After Michelangelo)

After thirteen billion years of radiant dance,

Orchestrated in ethereal harmony,

Pulsating with the quantum rhythms of the

Subatomic realm,

Another hand emerges.

Four fingers and an opposable thumb

Grasp into the air,

Out of the oceanic womb,

Reaching for life,

Craving validation for being,

Depending on Love for survival.

NOH TIME #2 (Front)

(After Michelangelo), 1985, 43" x 49," Wood, Oil/canvas,
Phillip Romero (Photographed by Jennifer Douglas)

NOH TIME #2 (Front)

(After Michelangelo), 1985, 43" x 49," Wood, Oil/canvas,
Phillip Romero (Photographed by Jennifer Douglas)

III

CREATIVE INTELLIGENCE TRAINNG

The Brain is a Creativity Machine

My First Meeting with Andy Warhol, 1976, David Richmond

"You have to treat the nothing as if it were something. Make something out of nothing."

Andy Warhol

The 6Rs of Creative Intelligence Training: *Reinventing Oneself*

Creative Intelligence Training © (CIT) mindfully integrates innate resilience to adversity with the brain's default creativity networks. With CIT we can mitigate the accumulation of everyday stresses escalating into chronic, toxic stress. Toxic stress produces *Burnout*, the breakdown of the stress regulatory system and inflammation.

Creative Intelligence Training (CIT) focuses mindful coordination practices to integrate genetically imprinted resilience with hard-wired brain-mind creativity. *Resilience + Creativity x Mindful Coordination = Creative Intelligence.* Creative Intelligence Training focuses *Logosoma Living* to amplify creative resilience.

The 6Rs of CIT focus the brain's attentional networks to *Remember, Reflect, Reframe, Reimagine, Reinvent, and Reconnect.* CIT applies these steps in three relationship domains, *mind-body, self-other, self-social system*. Regular practice of CIT rewires neural circuits to mitigate stress, to 're-parent' the body, and to overcome insecure attachments of the past. This process empowers the reinvention of an authentic self in relationships and social networks.

Creative Intelligence can be amplified, empowered, and shared with Creative Intelligence Training © (CIT). CIT cultivates a dynamic balance of different brain states for *being* and *doing*.

The Brain-Mind System: *A Creativity Machine*

The human brain-mind system experiences *consciousness*, self-awareness. From cradle to grave, the greatest threat to individual life is other humans. When relationships are insecure, chronic, toxic stress hijacks the brain and they live in perpetual *freeze, fight, flee* mode—very unstable. To activate the OFF SWITCH TO STRESS in the brain, we must employ mindful self-regulation. By calming the breathing, we activate the RELAXATION RESPONSE: *Calm, Connect, Create*. Now the brain can make assessments and conversations to defuse and redesign the relationship to more efficient coordination of action: Creativity gives rise to individual peace and prosperity, family longevity, and culture flourishes.

Mind-Body Attachment: *Being and Doing*

Awareness is perceived as *attachment* between two different categories—the mind and the body. The mind-body *relationship* is rooted in the emotional bonds between mother and infant at the beginning of every human life. Emotional bonds confer greater resilience for survival in mammals and other species. This bond can be assessed on a spectrum of secure attachment and insecure attachment. The mammalian body is designed to detect insecurity, and the brain triggers a stress response when it perceives a threat. Every attachment, from mind-body to self-other to self-society is influenced by the complex emotions of relationship stress. Mindful awareness of this phenomenon can mitigate insecure attachment and create new patterns in relationships that are more secure. Stress in relationships activates the innate creative resilience for adaptation and survival.

The brain-mind system can be viewed as a *creativity machine* that spontaneously makes assessments of itself in *relationship* to other people, to nature, and to the world. We use creativity to adapt to stressful experiences and to overcome adversity. With effortful attention, the brain-mind system can focus executive functions to perform mental and physical tasks.

The "doing brain" state activates *convergent thinking,* task oriented mental processes. When the mind choses to suspend executive focus and "stop doing", it activates *divergent thinking,* a state of "being aware" emerges.

Mindful awareness of these two distinct brain states, "doing" and "being," is critical to a cultivating a balanced mind-body system. Practicing *daydreaming* refocuses the busy executive brain-mind from point-focused, goal oriented, linear thinking and doing to nonlinear, free association, creative processes that we know as daydreaming. This state of wonderment is equivalent to the 'child-mind" of discovery, exploration, and experimentation of play. When daydreaming is attained, many brain centers of consciousness begin to communicate. The attentional networks of the brain turn inward, and the whole brain activity becomes synchronized in sharing information. Links between memory, emotion, the five senses, imagination, and problem solving mobilize the neural substrates of creativity. Cultivating daydreaming as a mindful practice empowers our creative nature in reinventing our sense-of-self, our attitudes, and our behaviors with each other.

Creative Intelligence: *Mindful Daydreaming and Autopoiesis*

Cultivating daydreaming as a mindful habit for mental and emotional hygiene can mitigate perceived stress, detoxify relationship stress, and active creative resilience. Creative Intelligence is critical to empowering the innate process of *autopoiesis,* self-making.

Autopoiesis is a fundamental phenomenon in living systems. The brain's neuroplasticity, the ability to learn new patterns of thinking and behavior, can be promoted with the practice creative resilience. This practice allows for continuous creative adaptation to life's experiences, positive and negative. By reinventing ourselves, we can improve our joy in living, promote longevity, and empower the evolution of more secure attachments in social systems from the nuclear family to the "global village".

Th 6Rs of Creative Intelligence

The practice of Creative Intelligence is divided into two categories of attentional focus: Mindful practices (abstract) and Mind-body practices (concrete). Mindful practices focus attention inward, minimizing body activity, to activate creative, divergent thinking and brain synchronization. Mind-Body practices, outer directed attention, drive creative patterns of behavior. The 6Rs of Creative Intelligence are:

Mindfulness Practices

1. Remember
2. Reflect
3. Reframe
4. Reimagine

Mind-Body Practices

5. Reinvent
6. Reconnect

3.1 Remember

Memory is the foundation of our sense of an authentic sense of *self*. Intrusive awareness of negative emotional memories, *phantom stress*, can increase their toxic effects and create symptoms of anxiety, depression, and self-sabotaging behaviors.

Remembering positive emotional experiences can increase our resilience to adversity and diminish the toxic effects of negative memories.

Reparenting the Sense-of-Self: Detoxifying negative emotional memories

The practice of focused attention on a single negative emotional memory is not pleasant. Psychotherapy provides a safe environment with a trusted professional to reflect on these memories. Reparenting is a mindful practice where the mind is the loving parent, and the body is the emotional child.

A mindful approach can be employed to contemplate the painful memory. This skill requires the development of the mental-emotional construct where the mind becomes a parental source of loving kindness reassure the body that it is safe. This requires considerable practice of self-compassion meditation. Like exposure therapy, where we confront painful trigger experiences to desensitize the stress reaction, mindful remembering reawakens the feeling from pain, fear, confusion, resentment, sadness, self-pity, and much more. With gradual excursions into these memories, the brain detoxifies the negative emotions, depositing them as past, harmless events.

3.2 Reflect

Mindful reflection on negative emotional memories allows for objective understanding and insight. We can *accept* that the body can *feel* like a victim or hostage when adversity occurs as a part of life. We can distinguish between "there-and-then" memories from the past from "here-and-now" experiences in the present. With *gratitude* for our mind's creative ability, we can adapt to and detoxify the negative circumstances. When we mindfully reflect on negative memories, they are re-remembered. Neuroscientists call this *the reconsolidation of memory*. Negative emotional memories are literally rewired, unlearned, and transformed from "high negativity" to "low negativity". Trauma and chronic stress memories are rendered as harmless events in the past. This is how psychotherapy works.

3.3 Reframe

Reframing the feeling of being a victim or hostage is a creative, mindful practice. By declaring, "I am not a victim" the mind functions like a protective parent to the emotional body, which is like a child. This reframing practice functions as a "declaration of independence" from negative emotions, the first step in creative resilience to adversity.

Learning to reframe stress triggers requires a commitment to daily practice. The Mindfulness Script and the 4Rs of Logosoma Brain Training demonstrate how to accomplish this.

3.4 Reimagine

Reimagining changes the brain state of "doing" the behaviors of everyday life, *convergent thinking*, to the brain state of "being", *divergent thinking*. The brain is a creativity machine. When the mind refocuses attention to "stops doing" and focuses on "just being," attention turns inward, as in daydreaming or meditation. Non-linear patterns of awareness emerge, and the mind wanders aimlessly, exploring awareness from multiple brain centers of consciousness. These neural networks spontaneously and non-consciously synchronize in their electrochemical communication. They begin communication with each other, activating creativity, and inventing new ways to solve problems. The observing mind, paying attention this synchronized "brain chatter", may experience insights and solutions to problems during this process.

Mind-Body Practices

3.5 Reinvent

Reinventing is an active, mind-body, creative process. It is empowered by documenting the new patterns of thinking and acting. Through writing, drawing, and researching, we can reinvent new attitudes and behaviors. Journaling is helpful in the exploring, inventing, risk taking, and practicing this step. Children are actively engaged in this process from the earliest days of life.

3.6 Reconnect

The primary reconnection to focus on is the mind-body connection. Emotional intelligence can be increased with this new brain circuit training. A heightened awareness of the self and the social networks we live in reduces relationship stress. With this reconnection process of reinventing the self, you become ready to reinvent your social connectivity for deeper, meaningful bonds.

Shared Creative Intelligence is empowered with mindful social communication practices in other Logosoma Brain Training techniques.

Empowering Shared Creative Intelligence: *A Family Affair*

Our Shared Creative Intelligence begins in families during toddlerhood. Families are the core unit of culture—like the single cell of complex organisms. When families create secure emotional bonds with shared creativity everyone thrives. Families and individuals with emotional intelligence and creative intelligence empower creative contributions to society. Shared Creative Intelligence reciprocates in the family-culture system, empowering social flexibility toward a more egalitarian society.

Likewise, families are the crucible for attitudes and beliefs that undermine authentic identity development and social reciprocity. Using language to express negative emotions is critical to the development of emotional intelligence. Families that "act-out" negative emotion rather than express them with language develop alexithymia, the absence of emotional intelligent language. Domestic violence increases in alexithymia systems. Overvaluing obedience over spontaneous creativity in play contributes to a dissociation of the mind-body awareness and increases fear of authority and shame. Inadequate validation for 'just being,' and overvaluing 'doing it right' contributes to anxiety, perfectionism, and self-doubt.

Shared Creative Intelligence cultivates secure emotional bonds in families based on the development of an authentic individual sense-of-self in children. Feeling secure in the mind-body begins in the mother-infant bond and expands into the family network. CIT guides attachment with balanced being-doing attitudes and behaviors. The social bond within families confers wellbeing and longevity, expanding into a wish to make social systems more secure.

S.O.S.: *Shared Creative Intelligence and Species Tribalism*

As guardians of the self-organizing-system of human cultural evolution, we must focus on promoting Species Tribalism as a core value of the Global Village. This requires a collective application of the 6Rs of Creative Intelligence.

- *Remembering* the thousand-year patterns of human self-destruction reveals primitive social-emotional development. The atrocities perpetrated on one-another, driven by greed and fear, are rooted in the developmental cognition of 2-year-old toddlers who cannot take-turns and share.
- *Reflecting* on the painful past reveals that present human existence has never been better for so many people.
- *Reframing* the self-sabotaging patterns of our nature as a developmental phase that can be overcome with education inspires hope.
- *Reimaging* a world with cooperation, collaboration, and creativity, points the way toward a sustainable biosphere.
- *Reinventing ourselves* beyond the destructive co-dependency will require new collective methods of self-regulation.
- *Reconnecting* the Global Village with these skills promises a planet where all life is precious, and we are the guardians of the complex self-organizing-systems of life.

NOH TIME 1988/1400

Zeami's Noh Theater dramatized

The style of the flower,

Inviting the gaze at the fragile blooming,

Beaming with joy at radiant blossoming,

And weeping at the desiccation of petals.

Zeami invited mindful attention into

The perfection of time.

Quantum aesthetics permeate nature

And charge the creative forces of the present moment

With timeless wonder,

Beauty, and perpetual sadness.

Nature creates each moment

Entangled with destructiveness.

The cry of the infant's first breath

Marks the death of the mother's bliss in pregnancy.

The pain of living begins with clutching to life,

As mother clings to the desperate outcast

Pushed from her ocean of cosmic gestation.

NOH TIME #1

(for Zeami Motokiyo), 1986,
86" x 60" Oil/canvas, Phillip Romero (Photographed by Jennifer Douglas)

NOH TIME #1

(for Zeami Motokiyo), 1986,
86" x 60" Oil/canvas, Phillip Romero (Photographed by Jennifer Douglas)

IV

<u>PSYCHOPOIESIS TRAINING</u>

Psychedelic Medicine and Selfless Mind

Lao Tzu, Song Dynasty statue of Chinese master of Taoism. Located in Quanzhou.

"Great indeed is the sublimity of the Creative, to which all beings owe their beginning, and which permeates all heaven."

Lao Tzu

93

4.1 Psychedelic Medicine: *Rewiring the Sense-of-Self*

Psychedelic medicine brings the potential for amplifying neuroplasticity to rewire the sense-of-self—*psychopoiesis.* The critical human experience produced by psychedelic substances is the temporary *loss of the sense-of-self*, the ego. Lucid consciousness of timeless wonder emerges. Feelings of *oneness* occur, of being one with the universe—*Selfless Mind.* Experiencing awareness of *oneness* beyond the constraints of space-time, attachment, the dialectics of perception from life-death, good-evil, right-wrong, past-future, self-other is exhilarating.

Selfless Mind experience promotes the unity of emotional and cognitive processes that reflect the alteration in the brain's Default Mode Network (DMN), the resting sense-of-self. As a participant-observer of the complexity and synchrony of natures' processes, the selfless mind activates a deep sense of gratitude and willingness to accept the trials, tribulations, and adverse experiences of life. Acceptance of impermanence opens selfless mind to liberation from fear, anger, and confusion that arises with the attachment stress experienced by the ego. Psychedelic substances empower human Creative Intelligence with the potential to reinvent the sense-of-self.

At the neural network level, psychedelic substances activate the brain's neuroplasticity. Evidence attributes the therapeutic potency of psychedelics to this phenomenon. Psychiatric disorders, including Treatment Resistant Depression (TRD), posttraumatic stress disorder (PTSD), and others, appear to be associated with neural network dysfunction from toxic corticosteroid effects of stress. By activating the brain's ability to rewire itself, symptoms of depression, anxiety, and cognitive impairment are temporarily mitigated.

To promote long-term stability of the neuronal benefits, Logosoma Training provides a brain-based method of recruiting the frontal attentional networks for mindfulness. With focused attention the mind can narrate the recreation of the self. With practice, these changes can rewire the neural substrates of the sense-of-self toward greater resilience, peace, and joy in living.

4.2 Theory of Mind: *Despair vs. Depression*

Attitudes profoundly affect behavior, emotion, cognition, and socialization. When attitudes become fixed in negativity by a distorted sense-of self, a chronic state of despair hijacks the brain and cripples the mind. Creativity is paralyzed and personal rigidity impairs functioning. This is the burnout syndrome

Brain Scan #1: *Evolution of the Human Brain: From Matter to Mind Gerhard Roth, Ursula Dicke, in Progress in Brain Research, 2019*

Theory of mind

Theory of mind (ToM) is defined as the ability to understand and take into account *another individual's mental state* or of "mind-reading" (Premack and Woodruff, 1978). In children, the ability to implicitly understand the intentions as well as false beliefs of others is already present around the age of 1 year (Baron-Cohen et al., 2013; Meunier, 2017), while "full-blown" or "explicit" ToM and the understanding that a person can hold a false belief develops between the ages of 3 and 4 years and is fully developed only at the age of 5 (Flavell et al., 1978, 1981)

Using his Theory of Mind Scale, Jeremy Coplan distinguishes the sense-of-self from low self-esteem in Treatment Resistant Depression. Depression alters the sense-of-self, but when antidepressant treatments work, some people persist in having marked impairment in their sense-of-self and maintain attitudes of despair.

Brain Scan #2: *Depressive Response and the Rostral PFC: A Novel Assessment of Emotional 'Theory of Mind' Changes Following rTMS, Ryan Webler BA, Tarique Perera MD, Jeremy Coplan MD, SUNY Downstate University Hospital*

> Coplan concludes:
>
> -Non-response in our sample was associated with continued disturbances in self-reflection and perception of others as measured by the Coplan Emotional Theory of Mind Scale.
>
> *-As hypothesized, an individual's sense of self and how they are perceived by others, their 'emotional theory of mind', is distinguishable from (though related to) standard depressive symptoms measured by the BDI (Beck Depression Index)*
>
> -Emotional theory of mind was a treatment resistant symptom set in the context of rTMS (right transcranial magnetic stimulation) …

Creative Intelligence Training: *New the Sense-of-Self Script*

Neuroscience maps the circuits of the sense-of-self. Adverse childhood experience (ACE) and chronic, toxic stress can distort the neural networks of the sense-of-self. The consequences of such damage can be sense-of-self narrative charged with self-loathing, shame, guilt, self-sabotaging behaviors, blame, victimhood, and hopeless despair. Chronic toxic stress hijacks the brain and the sense-of-self becomes a hostage lifelong mental and emotional dysfunction, social isolation, and critical impairment of the innate capacity for creativity.

Creative Intelligence training uses these brain maps to focus mindful practices for redesigning the neural substrates the produce the dysfunctional narrative of the sense-of-self. With ketamine enhancement of CIT, the time to rewire the sense-of self can, theoretically, be significantly reduced. In the window of heightened neuroplasticity, a 'New Identity Script' provides a daily practice to enhance new neural networks and a renewed sense-of-self. Declaring this script out loud can amplify the changes rapidly.

- *I,   , am grateful for my mind's Creative Intelligence.*
- *I love my body with all the emotions it can feel.*
- *I am excited to create new ways to nurture, protect, and strengthen my body.*
- *I am excited to create new attitudes and behaviors for my mind-body sense-of-self.*
- *I look forward to practicing my new identity for wellness and longevity.*

4.3 Mastering the Anxiety of Impermanence: *Selfless Mind*

Buddha's Four Noble Truths identify to root cause of human suffering as the egoistic fear of loss of the natural attachment to life. The Truth of *Impermanence* guides mindfulness practices of focused attention, contemplation, moral attitudes, ethical behaviors, and meditation. These practices lead to the realization of non-attachment, the attainment of consciousness without an egoistic self—*selfless mind*. Deep acceptance of impermanence frees the conscious mind from suffering. A profound connectedness with the world emerges and primordial feelings of compassion and love for being floods consciousness.

Psychedelic substances trigger this glimpse beyond the clinging sense-of-self with the temporary disruption of the DMN. Selfless Mind experiences hold the potential to change the suffering, ego-centric attitudes and behaviors that drive human behavior. Fear drives self-sabotaging patterns of thought, emotion, and action toward greed, competition for control, struggle for power, and conflict.

With the deep acceptance of impermanence, the anxieties triggered by aging, loss, illness, and death dissolve. Real joy in the perpetual changes in life can be profoundly satisfying, and the capacity to cultivate wider compassion nourishes shared Creative Intelligence.

NOH TIME #3 (Back)

Back, 1992, Japanese lacquer on wood, bamboo, washi, gold leaf,
26" x 28" Phillip Romero (Photographed by Jennifer Douglas)

NOH TIME #3 (Back)

Back, 1992, Japanese lacquer on wood, bamboo, washi, gold leaf,
26" x 28" Phillip Romero (Photographed by Jennifer Douglas)

PRISONERS OF TIME

Primordial Directive

Freeze, fight, flee.

The brain is hijacked by fear of pain,

Terrified for its life,

And hardwired to survive.

Anger, fear, and confusion cloud consciousness.

Rational thought is unplugged to mitigate harm

And promote the chances for life.

In a hormone storm,

The body awakens its primeval script,

Etched in the DNA,

And mapped in the complex machinery

Of the body.

Forfeiting all learned patterns of living

Acquired over two million years of evolution,

The mind-body link fractures.

No mind, no self,

No tomorrow, no yesterday,

Only now, and now only,

Life faces extinction.

Toxic Fear

The infant's joy, wonderment, curiosity, and play,

Utterly depend on Mother's holding presence.

Omnipotence pervades consciousness of the body-self,

Empowered by mother's attuned care.

Terror sweeps the fragile balance

Between mother and child

When trauma, great and small,

Disrupt the secure bonds of symbiosis.

Toxic fear takes the brain-mind-self hostage.

Soaked in a poisonous flood of cortisol,

The sense-of-self lives in fear,

Constantly on high alert,

Unable to free oneself from

The perception of threat.

The hostage mind-body-self

Cascades into self-sabotaging behavior.

Addiction, self-harm, domestic violence, gambling with life...

Trapped in a victim-script,

The sense-of-self spirals toward resignation.

V

<u>CHANGE the SENSE-OF-SELF;
REMAKE the WORLD</u>

Red Self-portrait, Leonardo da Vinci, 1512, <u>Courtesy of LeonardoDaVinci.net</u>

"Principles for the Development of a Complete Mind: Study the science of art. Study the art of science. Develop your senses- especially learn how to see. Realize that everything connects to everything else."

Leonardo Da Vinci

5.1 Individual Change and Cultural Evolution: *Complex Systems*

The primary mission of Logosoma Living Wellness Protocol is to recruit the Creative Intelligence to rewire the neural networks of sense-of-self. With practice, the skills can empower autonomy, resilience to stress, and create secure bonds in families and social systems. With psychedelic enhanced Psychopoiesis Training, consciousness taps into its hardwired mammalian networks for compassion. This experience of cosmic awareness and love of living forms the basis for the use of psychedelics to promote social change. Increased awareness of connectivity for all humankind empowers communication, cooperation, and collaboration for creating improved ways to live together.

5.2 Species Tribalism: *The Cultural Prime Directive*

Mammalian survival depends on working with others as critical to success. The pattens of behavior in all mammalian species share a 'culture of connectedness'—*survive together, thrive together.* This implicit 'prime directive' is encoded in the mammalian genome.

Brain Scan #1: *Cultural evolutionary theory: How culture evolves and why it matters, Nicole Creanzaa,1, Oren Kolodnyb,1,2, and Marcus W. Feldman, Department of Biological Sciences, Vanderbilt University, Nashville, TN 37235; and Department of Biology, Stanford University, Stanford, CA 94305, 2017*

Human cultural traits—behaviors, ideas, and technologies that can be learned from other individuals—can exhibit complex patterns of transmission and evolution, and researchers have developed theoretical models, both verbal and mathematical, to facilitate our understanding of these patterns… Furthermore, cultural, and genetic evolution can interact with one another and influence both transmission and selection. This interaction requires theoretical treatments of gene–culture coevolution… in addition to purely cultural evolution… the core concepts in cultural evolutionary theory… pertain to the *extension of biology through culture*… the societal implications of the study of cultural evolution and of the interactions of humans with one another and with their environment.

Neolithic humans, Cro-Magnon, were challenged by nature to survive for 300,000 years. They initiated modern culture 40,000 years ago with the creation of cave art—one of the first signs of Creative Intelligence and mindfulness. By creating symbolic images, the brain-mind system took a great leap forward in creating an abstract concept of humankind in nature. Our ancestors could now use art to stimulate their imagination. The process of cultural evolution began the reinvention of how to live with each other. Further leaps in observing nature's patterns produced farming and the creation of the tribal village. The domestication of plants and animals for food generated a demand for new social hierarchies, labor specialization, information gathering, trade, monetary and military systems. With the ability to store food and goods, the power of village life kindled the accumulation of excess.

The emergence of individual and social greed rapidly expanded. Greed sparked human conflict and amplified the need for military culture. As humans domesticated food and energy sources, the primary threat to survival was not nature but other humans.

Creative Intelligence charged both creative technologies and the destructive use of these technologies against each other. The mission to create sustainable cultures would drive human history toward expansion and conflict. Our present-day global village teeters on the precipice of an unsustainable biosphere. Once again, we must reinvent how we live together if we are to maintain a functional civilization and a vital biosphere.

5.3 Beyond Global Burnout: *A Creative Intelligence Imperative*

Crisis kindles cultural evolution, and the deep instincts for survival recruit cooperation in mammals. I reflect on this process in *Global Village Burnout: Creative Intelligence = Survival* (2021). Multiple, complex existential crises challenge human Creative Intelligence as never before. The ability to collectively address these issues is paralyzed with toxic global stress. It is as if we are on a race to extinction.

Human beings generate belief systems that impair and impede survival instincts. Fear and greed drive humans to commit atrocities against each other. As we struggle to learn how to live together in an internet-connected global village, we are witnessing collective regulatory systems breakdown as insecure social system stress escalates. A pandemic of escalating toxic stress is spreading through social media. Socioeconomic divisiveness is rampant.

A biopsychosocial approach comparing patterns leading to global system burnout with patterns in family systems hijacked by addiction may be helpful. Family systems form the core unit of culture like the single cell forms the basic unit of the human being. With understanding the mechanisms of cellular biology, we can upscale designs for wellness practices for the entire organism.

Addiction, Relationship Stress, and Burnout: *Family Systems*

Addiction is a compulsive pattern of self-sabotage in individuals and families. Based on predictable consequences of chronic stress, the problem is often rationalized, minimized, and denied by the addict and the family—the triad of codependency. If we examine the natural breakdown of family systems with addiction, we may be able to scale up the effective steps of recovery for creative solutions to larger social systems.

Patterns in families struggling with addiction are well known. For over thirty years, Logosoma theory has informed successful interventions and recovery for individuals and families. The critical features of family systems with addiction include:

- chronic stress takes the individual and family system hostage
- rationalizing, minimizing, denying the addiction
- codependent attachment
- repetitious narratives of victim-victimizer-rescuer
- blame
- shame
- resentment and self-pity
- self-sabotage attitudes and behaviors
- greed over compassion
- burnout, breakdown

Rewiring the Sense-of-Self: *Remaking Family Systems*

Logosoma training identifies the patterns of addiction and codependency and guides mindfulness practices in a twofold process of recovery and reinvention. Simultaneous unlearning and relearning the neural networks that drive these pattens is structured in the exercises and practices.

Recovery from addiction requires changes in attitude and behavior for all members of the system. Among the essential changes are:

- Acceptance of the adversities of life
- Acceptance that one cannot control external events or individuals
- Gratitude for the neuroplasticity of the brain-mind system
- Gratitude that one can change the reaction to adversity
- Gratitude for the power of mind to self-regulate emotions
- Gratitude for Creative Intelligence
- Motivation to change one's attitudes and behaviors
- Stop blaming
- Stop shaming
- Self-compassion
- Self-validation
- Validation and empathy for others
- Willingness to co-create new patterns of connectivity
- Willingness to practice new patterns for secure attachment, cooperation, and collaboration

Learning and practicing the principles of Logosoma training support brain changes to reinvent the sense-of-self. As individuals change family system stress is mitigated and the possibility for reconciliation and reinvention emerges. With shared Creative Intelligence the system can create new patterns of living together. With maintenance of emotional communication, conflict resolution, and creative problem solving, new memories reshape the present into a more joyful experience for all.

5.4 The Challenge to Shared Creative Intelligence

Upscaling the practices that help family systems navigate stress may mitigate stress of larger social systems. The global village is facing numerous complex existential challenges. New technologies will not be adequate to change the present patterns in culture. Cultural evolution will require a paradigm shift in socioeconomic systems.

Like the Black Plague that triggered the end of feudal social structures and ushered in humanism and the Renaissance, our present-day pandemic is one of many stressors contributing to cultural breakdown. Shared Creative Intelligence must reinvigorate our humanistic traits to solve the challenges to the human systems that create the global village.

New social systems must integrate biopsychosocial principles for a sustainable biosphere. The consumer culture of modern era has become an addictive culture where human greed for energy, wealth, and power drive self-sabotaging patterns. The motivation to endlessly amass wealth is not self-sustaining. Sustainability and sharing must replace greed for growth as the defining core of world systems.

Implementing the great power of shared Creative Intelligence to explore possible new blueprints for human social systems is imperative.

REFERENCES

Abraham, Anna. *The Cambridge Handbook of the Imagination*. Cambridge University Press, 2020.

Abraham, Anna. *The Neuroscience of Creativity*. Cambridge University Press, 2019.

Andreasen, Nancy C. *The Creating Brain: The Neuroscience of Genius*. New York: Plume Book, 2006. Print.

Erikson, Erik H. *Childhood and Society*. London: Vintage Digital, 2014. Print.

Jung, Carl Gustav. *Carl Gustav Jung / the Structure and Dynamics of the Psyche*. Routledge, 1992. Print.

Kandel, Eric R. *The Age of Insight: The Quest to Understand the Unconscious in Art, Mind, and Brain: From Vienna 1900 to the Present*. Random House, 2011. Print.

Kokkinou, M., A. H. Ashok, and O. D. Howes. "The Effects of Ketamine on Dopaminergic Function: Meta-analysis and Review of the Implications for Neuropsychiatric Disorders." *Molecular Psychiatry* 23.1 (2017): 59-69. Print.

Kokkinou, Michelle, and Oliver Howes. "O7.1. Midbrain Dopamine Neuron Activity Controls the Effects of Repeated Ketamine on

Striatal Dopaminergic Function." *Schizophrenia Bulletin* 44. Suppl_1 (2018). Print.

Lou, H.C., J.P. Changeux, and A. Rosenstand. "Towards a Cognitive Neuroscience of Self-awareness." *Neuroscience & Biobehavioral Reviews* 83 (2017): 765-73. Print.

Margulis, Lynn, and Dorion Sagan. *What Is Life?* U of California, 2000. Print.

Maturana, Humberto R., and Francisco J. Varela. "Autopoiesis and Cognition." *Boston Studies in the Philosophy and History of Science* (1980). Print.

McEwen, Bruce S., and Harold M. Schmeck. *The Hostage Brain*. Rockefeller UP, 1994. Print.

McFarland Solomon, Hester. "Self-Creation and the Limitless Void of Dissociation: The 'as If' Personality." *Journal of Analytical Psychology* 49.5 (2004): 635-56. Print.

Rizzolatti, Giacomo, Frances Anderson, and Corrado, Sinigaglia. *Mirrors in the Brain: How Our Minds Share Actions and Emotions*. Oxford UP, 2008. Print.

Zacharias N; Musso F; Müller F; Lammers F; Saleh A; London M; de Boer P; Winterer G; "Ketamine Effects on Default Mode Network Activity and Vigilance: A Randomized, Placebo-controlled Crossover Simultaneous FMRI/EEG Study." *Human*

Brain Mapping. U.S. National Library of Medicine. Web. 23 June 2022.

Peter Pressman, MD. "The Role of an Fmri in Monitoring Brain Activity." *Verywell Health*. Verywell Health, 29 Jan. 2020. Web. 24 June 2022.

Dr. Romero's latest book expertly and succinctly weaves his teachings on psychiatry, Creative Intelligence and the four noble truths of Buddhism into a must read for clinicians, patients, and every day people who are looking to improve their lives. This seminal work not only sheds light on how our thinking mind can create un-ease and dis-ease, but also provides vital step-by-step training and a series realistic building blocks to reinvent or remake our sense of self.

- Jay Godfrey
  Co-Founder
  NUSHAMA Psychedelic Wellness Center, New York

Printed in the United States
by Baker & Taylor Publisher Services